MW01041510

< GET CODING! >

Get Coding with LEGO WeDo™

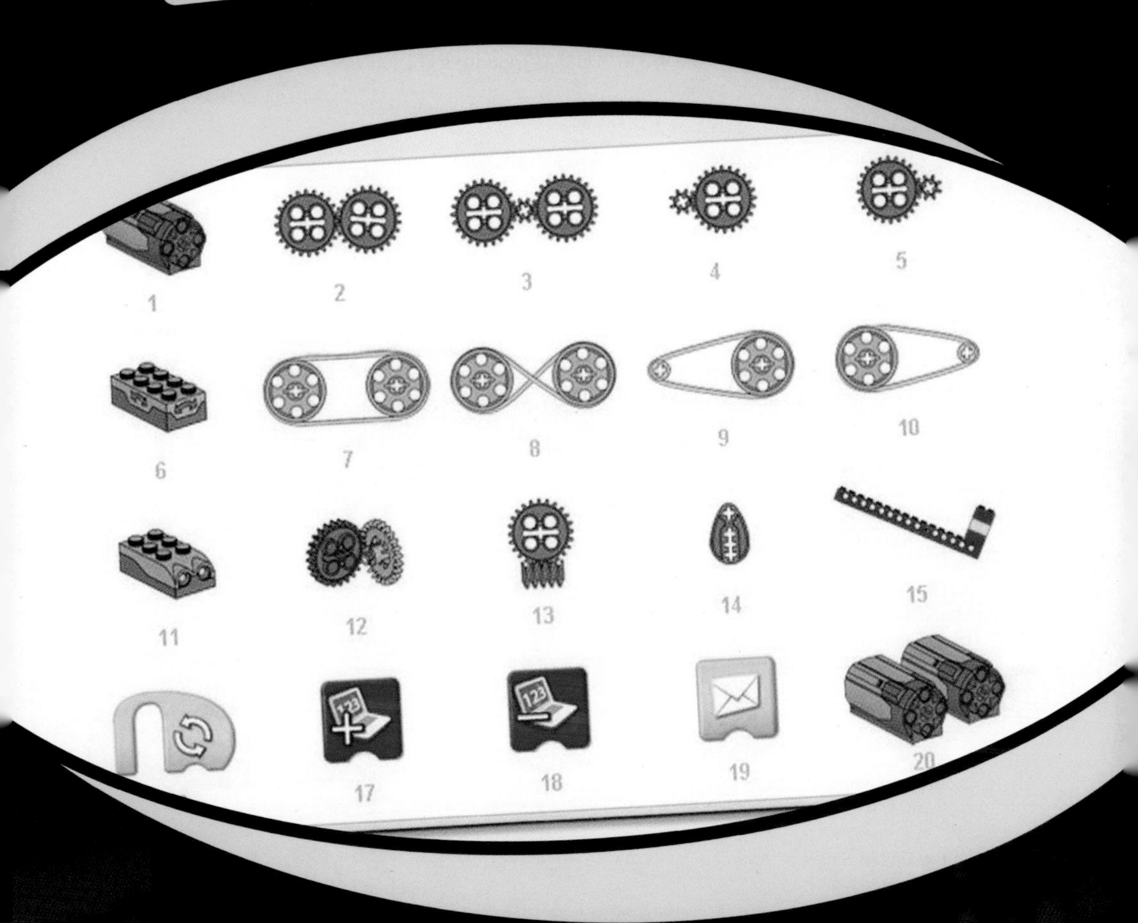

By Jenna Vale

Gareth Stevens
PUBLISHING

Please visit our website, www.garethstevens.com. For a free color catalog of all our high-quality books, call toll free 1-800-542-2595 or fax 1-877-542-2596.

Library of Congress Cataloging-in-Publication Data
Names: Vale, Jenna, author.
Title: Get coding with LEGO WeDo / Jenna Vale.
Description: Buffalo, New York : Gareth Stevens Publishing, [2024] | Series: Get coding! | Includes bibliographical references and index. | Audience: Grades 2-3
Identifiers: LCCN 2023007389 (print) | LCCN 2023007390 (ebook) | ISBN 9781538288580 (library binding) | ISBN 9781538288573 (paperback) | ISBN 9781538288597 (ebook)
Subjects: LCSH: Programming languages (Electronic computers)–Juvenile literature. | LEGO toys–Juvenile literature.
Classification: LCC QA76.52 .V35 2024 (print) | LCC QA76.52 (ebook) | DDC 005.13–dc23/eng/20230405
LC record available at https://lccn.loc.gov/2023007389
LC ebook record available at https://lccn.loc.gov/2023007390

Published in 2024 by
Gareth Stevens Publishing
2544 Clinton Street
Buffalo, NY 14224

Portions of this work were originally authored by Patricia Harris and published as *Understanding Coding with LEGO WeDo*. All new material in this edition authored by Jenna Vale.

Designer: Leslie Taylor
Editor: Megan Kellerman

Photo credits: Cover (illustrations) StonePictures/Shuttertstock.com; cover & series art (coding background) BEST-BACKGROUNDS/Shutterstock.com; (LEGO screenshots and graphics) courtesy of the LEGO Group; p. 5 Triff/Shutterstock.com; p. 11 Polytechnic Museum/Flickr.com; p. 17 (insert) Klaus-Dieter Keller/https://commons.wikimedia.org/wiki/File:Lego_WeDo_2.0_Glowing_Snail.jpg; p. 19 Klaus-Dieter Keller/https://commons.wikimedia.org/wiki/File:Lego_WeDo_2.0_Milo_part_B.jpg; p. 21 Joaquin Corbalan P/Shutterstock.com; p. 23 Brad Flickinger/Flickr; p. 27 Rizkan Yazid/Shutterstock.com; p. 29 AlesiaKan/Shutterstock.com.

Printed in the United States of America

CPSIA compliance information: Batch #CS24GS: For further information contact Gareth Stevens at 1-800-542-2595.

CONTENTS

Words in the glossary appear in **bold** the first time they are used in the text.

Get the Gears Turning

Kids have been building fun, colorful models with LEGO sets since the 1950s. From connecting block sets to programmable MINDSTORMS robots, the LEGO Group has created sets for almost anything you can think of! In 2009, the LEGO Group decided to add a new educational set to their robotics line: LEGO WeDo.

This set is all about learning the basics of engineering and coding. Builders create simple robots and then code programs to move them around using motors and gears. In 2016, LEGO upgraded to WeDo 2.0, which increased the number of elements in a set and offered wireless connectivity. WeDo makes it easy to dive into your first coding and robotics project!

This is Perseverance, a rover, or kind of space robot, that landed on Mars in 2021. You can make your own WeDo rover, but on a much smaller scale!

The Engineering Design Process

One of the most important ideas in engineering is the engineering **design** process. This is a **systematic** approach to establishing and solving a problem. Engineers test multiple approaches to see what the best answer is. It's all about trying out different things, making discoveries, and finding creative answers. That's just what you'll do with WeDo!

Building Blocks

The first step of creating with WeDo is to organize the LEGO elements in your kit. This will make it easier to find what you need when building, whether you're following a guided project or creating your own personalized model. The WeDo kit includes motors, **sensors**, gears, and other building pieces, which you'll organize in the kit tray by function.

The WeDo kit also includes colorful LEGO blocks and even decorative parts to give your robot personality! Your creation will be able to move, change speed, turn, and stop. These parts include gears, wheels, bands, and more.

The components shown below are some of the parts you'll need to make your robot come to life.

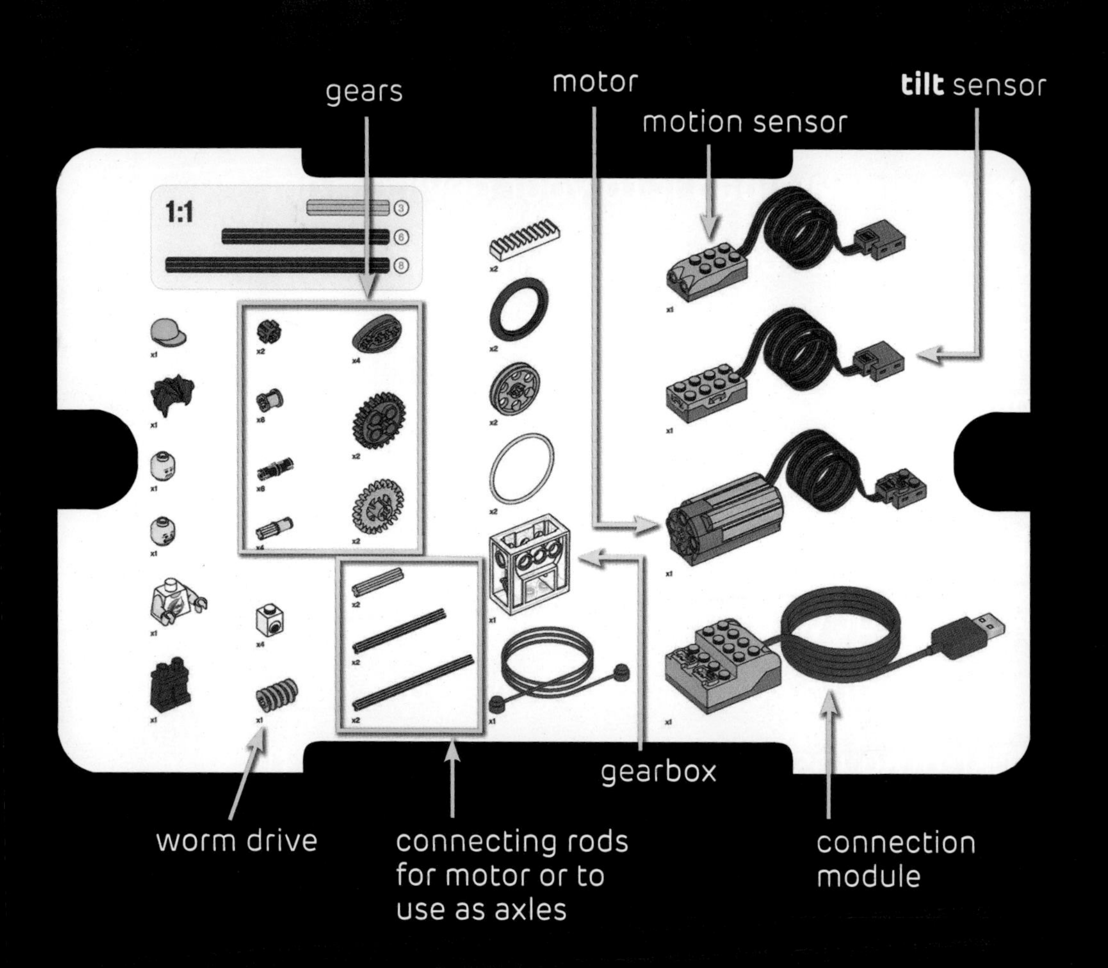

Get Your Ball Bearings

A new kind of part is also included with WeDo 2.0: ball-based elements. These allow your creations to move in ways they couldn't with WeDo 1.0. Milo the science rover, for example, has a rotating neck that you create by connecting parts with a ball bearing.

The WeDo Software

To get the WeDo 2.0 **software**, you'll have to **download** it to your device from www.education.lego.com. The WeDo **programming language** is very much like the blocks it controls. It features colorful "blocks" that coders drag and drop into a work area. These blocks fit together to make actions for your robot to perform.

Writing code in LEGO WeDo is simple. Just drag instruction blocks into a line to make your creations do what you want them to do. LEGO includes many projects in the kit that you can explore when you open the WeDo software.

This is the page you'll see when you first open the WeDo software. The robot seen here is the glowing snail!

Machine Automation

You'll write code that allows you to **automate** LEGO creations. The easy-to-understand WeDo software is a fun way to learn basic coding concepts. The colorful LEGO blocks, wheels, motors, and sensors help you understand simple robotics principles. Combining the two elements teaches important ideas about robotics and coding in real life.

Before You Code

It's important to learn the rules of coding before you begin programming in any language. This is a lot like learning the rules before you play a game.

Rule 1: Coders must know what they want the computer to do and write a plan.

Rule 2: Coders must use special words to have the computer take input, make choices, and take action.

Rule 3: Coders need to think about what actions can be put into a group.

Rule 4: Coders need to use **logic** with AND, OR, NOT, and other logic statements as key words.

Rule 5: Coders must explore the **environment** and understand how it works.

Learning how to code will help you learn problem-solving and computational thinking skills. These skills will help you with WeDo and beyond!

Applying the Rules

First choose a project from the WeDo kit or think one up yourself. Then you need to build the robot and think about what you want it to do and how you might get it to do that task. You must do these steps before you're ready to start coding.

The Coding Environment

The WeDo programming environment is where you'll write the code for your robots. Whether you're following along with one of the existing projects or doing your own from scratch, the software will always open with a Start Block waiting for you on the canvas, or workspace.

Notice the red square inside the yellow button at the bottom right of the screen. It stops the program if you make one that just keeps running all the time. The blue button at the top right of the screen is for **Bluetooth**, which you'll use to connect the software to your robot so it can perform the program you wrote.

The programming canvas is your workspace for LEGO WeDo. You'll drag and drop coding blocks onto the canvas to write your code.

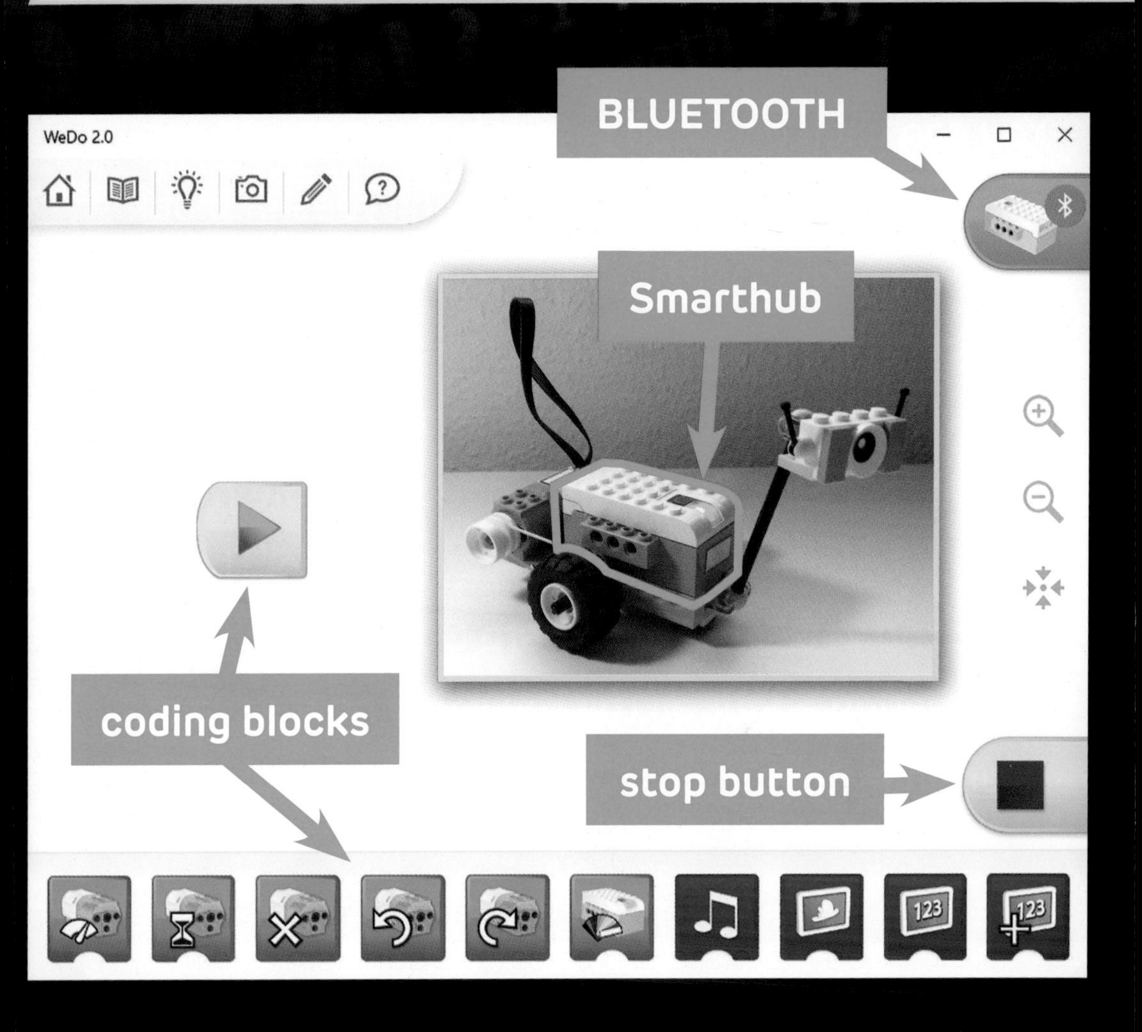

The WeDo Smarthub

The Smarthub serves as the brain of your LEGO WeDo creations. It allows the software to wirelessly communicate with any robot you build so you can see your code come to life! The Smarthub gets its power from two AA batteries or a separate rechargeable battery pack.

All the programming blocks are located in the menu, or list, at the bottom of the canvas. You can click or swipe through the menu to see all the options. There are six main types of block you'll be coding with, plus a comment block that doesn't affect your code.

If you want to learn more about any individual block, select the speech bubble from the toolbar at the top left of your screen. This will bring you to the Software Guide, which has information about all the tools you now have at your fingertips.

Comment Blocks

The Comment Block is a special kind of block that doesn't affect your code. It's used for leaving notes saying what each step of your code should be doing. The Comment Block looks like a speech bubble, and you can drag and drop it to the corner of any block you want to leave a note about.

The yellow, green, and red blocks connect from left to right. Blue and orange blocks connect to the bottom of the other blocks—see how they're shaped a bit like LEGO blocks?

Flow Blocks are used to control when and how the program runs. Use these blocks to establish when your program starts, when the robot should wait for something to happen, or when you need to create a **loop** in your code.

Motor Blocks are used to code how your robot's motor functions. Use these blocks to code how the motor turns, how long the motor is on, and even how strong you want the motor's power to be!

A Numeric or Text Input Block lets you add numeric or text values to any block you connect it to.

Light Blocks are used to light up the Smarthub's **LED**. There are 10 color options for the light, which can be changed by numeric, or number, input.

A Sound Block allows your program to make a sound with your device. A Display Block allows you to show images and text, or words, on the display of your device while your program is running.

A Sensors Input Block allows you to code behaviors for the tilt and motion sensors on your robot.

Let's Get Started

You're ready to get coding! Remember the glowing snail from page 9? It's one of the easiest robots to build and code with WeDo. From the software home page, choose "Your first project" to find the glowing snail activity. From here, the software will guide you through how to build and code the snail.

The code to make the snail glow is simple: just connect one Light Block to the Start Block and select the color you want the LED light on the snail's shell to turn. Experiment with different colors and additional Light Blocks to see what other effects you can create.

The software will guide you through this project from start to finish so you can get comfortable with building and coding.

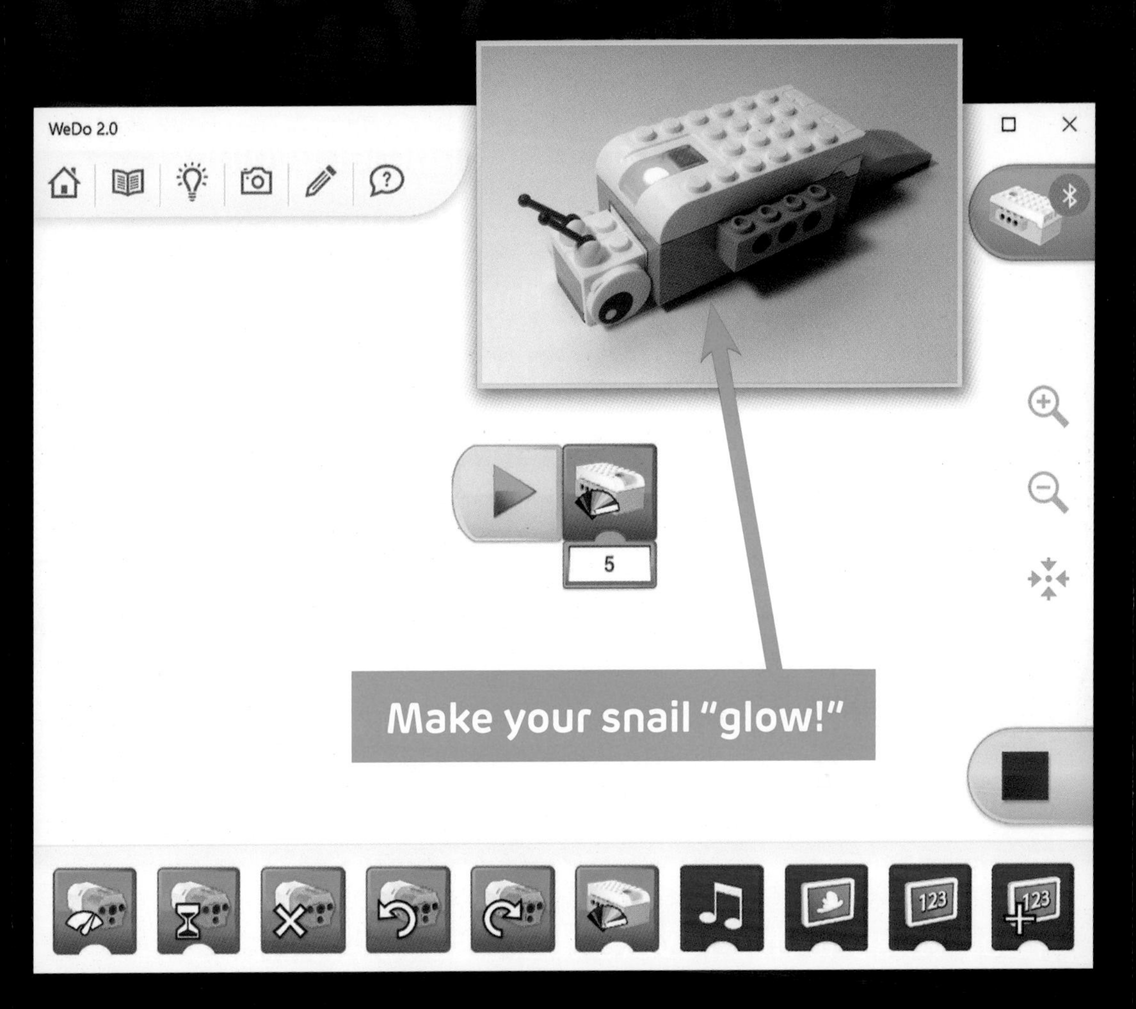

Check Out the Library

The glowing snail is a beginner project from the Project Library, which can be found by selecting the book icon from the toolbar. The library has beginner, medium, and advanced activities. You can also view the Design Library for base model robots to use as inspiration for making your own creations.

Meet Milo

Milo the science rover is another beginner robot you can build. While the glowing snail shows what programming the LED Block is all about, Milo has a motor and wheels. Working with Milo will allow you to experiment with programming movement.

Rovers are robots that can be controlled remotely, or from afar. They can even be programmed to move independently of human input thanks to artificial intelligence software. Today, NASA uses rovers to explore Mars! They've sent five rovers there so far. With Milo, you'll gain a better understanding of what scientists do when they want to explore somewhere they can't reach themselves.

Milo is just one model you can build, program, and experiment with.

Artificial Intelligence

Mars rovers now use artificial intelligence, or AI, to help them discover and study points of interest on the planet's surface. This means the rovers can copy human behavior according to a narrow set of guidelines. AI allows rovers to make discoveries without as much human input.

Coding Movement

You'll use the green motor blocks to code movement for Milo. The first kind you should add to the program is the Motor Power Block. This block says how fast the motor will spin. Set the power with a number from 0 (not spinning at all) to 10 (top speed!).

Next, decide which direction the motor will turn in. Choose the Motor That Way Block to make Milo move forward. This block has an arrow on it pointing in a clockwise direction. If you tap the block once it's part of your code, it will change the arrow direction. This would make Milo move backward.

Coding Milo's speed and direction will teach you the basics of programming any WeDo model that uses a motor.

Motion and Tilt Sensors

Have you ever activated the motion detector on an outdoor light? The WeDo motion sensor functions the same way, except instead of turning on a light, it starts the next part of your program. The tilt sensor starts the next part of the program when the item it's connected to is tilted in a direction you choose.

The next block you'll add to Milo's code is a Motor On For Block. This block says how long the motor will run for. Decide how long Milo's motor should run and input a number from 0 (0 seconds) to 10 (10 seconds).

The final block of code you'll drag and drop into the series is the Motor Off Block. This block will make Milo stop moving. You've completed your first program for Milo! The Motor Off Block is the final block for the program you've just written, but you can still add more blocks if you want to keep experimenting with Milo.

Do More with Milo

Try adding a Sound Block to Milo's code. The sound plays from your device, and you can test the block by choosing different numerical inputs. Each number is a different sound, and there are 28 options. Sound No. 1 sounds like it was made for Milo—try it out to see if you agree.

This is what the final code for Milo's program will look like. Try different motor speeds, lengths of time, and adding more blocks to your own program.

Follow the LEGO Brick Road

You can build upon simple programs like the one for Milo's movement in two ways. One way is to keep adding blocks to the line of code you've already written. This will result in one line of code for your program, and the actions will be performed one after the other. This is called a linear sequence.

You can also add additional lines of code near your first line on the canvas. If you make two separate lines of code next to or on top of each other, you can make your program carry out multiple actions at the same time. This is called a parallel sequence.

The envelope blocks attached to the Start Block send messages to the two parallel, or side-by-side, lines of code so they begin at the same time.

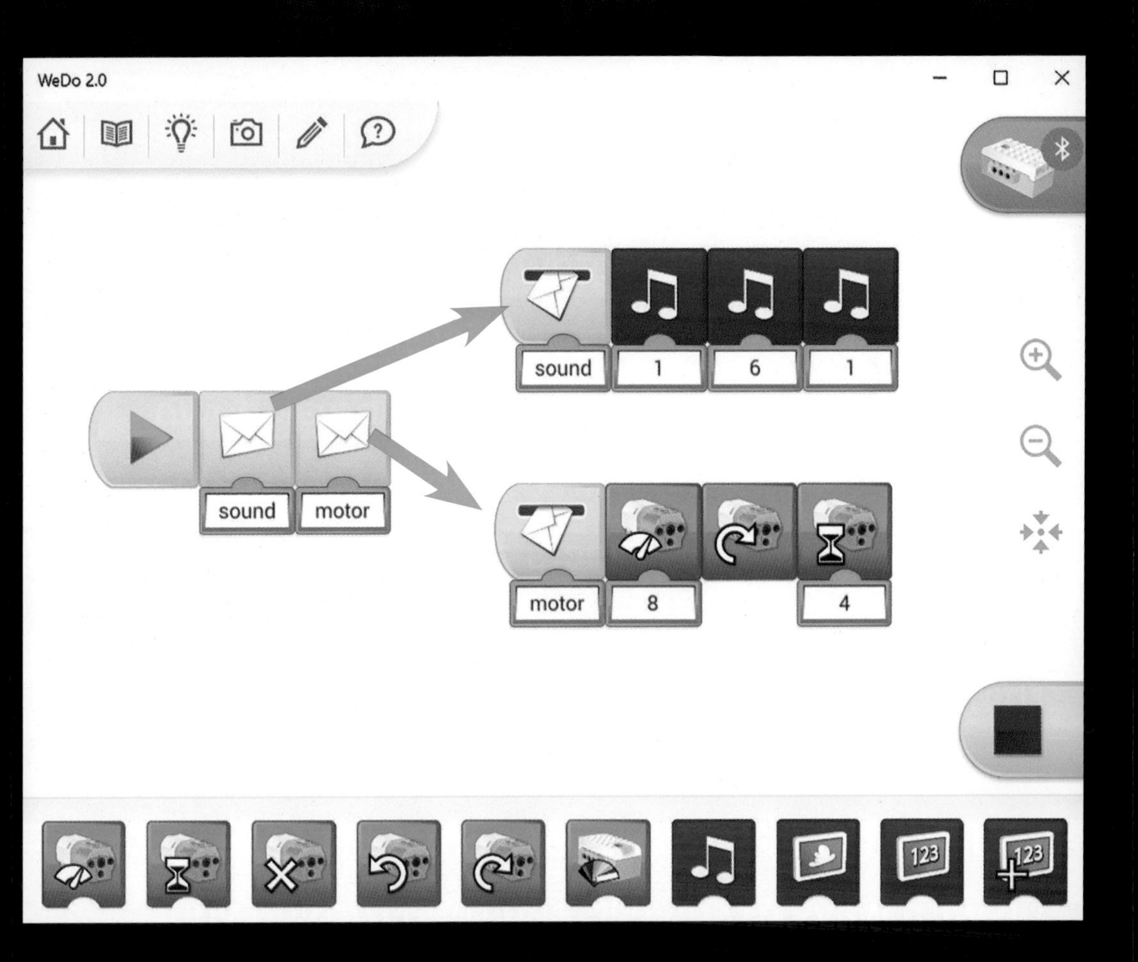

In the Loop

If your linear sequence is getting long, try studying it for any actions that can be grouped and looped. The yellow Repeat Block is a kind of Flow Block that can help you with this. Put the code for the actions you want to repeat a certain number of times inside the Repeat Block.

Computational Thinking

The more coding projects you do with WeDo, the more you'll grow your computational thinking skills. These are special skills for problem-solving that can help you break down harder problems and gain the ability to tackle them. Many WeDo projects are about answering real scientific questions, and computational thinking is key.

In coding, computational thinking means understanding the parts of a problem, coming up with ways to solve it, and then turning your ideas into code. It helps you sharpen your code so it's as simple, effective, and organized as possible. Using every part of computational thinking will help you succeed.

These are the four main components, or parts, of computational thinking. You can apply these skills to coding and many other kinds of problems in daily life.

Break It Down

decomposition
break the problem down into smaller parts

pattern recognition
seek out similarities between the smaller parts

abstraction
determine what information is most important

algorithmic thinking
write a series of rules or steps to solve the problem

Computational Thinking
create solutions that a person and a computer can understand

Step by Step

WeDo projects are made up of stages to help you improve your computational thinking skills. You might notice when you're doing projects that the software moves through three stages called Explore, Create, and Share. These stages help you learn to apply each part of computational thinking to the problem at hand.

WeDo and Beyond

WeDo has so many guided projects and activities for you to build, code, and learn with. But when you're ready to move on to harder models and code, there are other options you can explore. There are the MINDSTORMS EV3 and Robot Inventor sets, but if those seem a little too tough, try the SPIKE Prime set.

SPIKE Prime is a coding and robotics set that builds on what you've learned with WeDo. The coding system is slightly more advanced and there are more LEGO elements to work with. Keep building and coding. As you've learned with WeDo, there's nothing you can't do!

This is one of the models you can build with SPIKE Prime. What kind of code would you write for it?

Learning a New Language

You can do more with WeDo by programming your creations in another language called Scratch. This is like the block coding you've already been doing, but more advanced. You can connect software for Scratch directly to your creations. SPIKE Prime and MINDSTORMS also use block coding like Scratch.

GLOSSARY

automate: To cause something to run on its own.

Bluetooth: A standard for connecting devices within a short range of each other wirelessly.

design: A plan for a project.

download: To move or copy a file or program from one computer or device to another.

environment: The combination of computer hardware and software that allows a user to perform various tasks.

LED: A light-emitting diode. Also, an indicator light on an electronic device.

logic: A proper or reasonable way of thinking about or understanding something.

loop: A sequence of instructions that repeats until a certain condition is met.

programming language: A computer language designed to give instructions to a computer.

sensor: A device that reacts to a change in the world around it and then performs a task.

software: A program that runs on a computer and performs certain tasks.

systematic: Relating to a set plan or procedure.

tilt: To make something lean in another direction.

FOR MORE INFORMATION

Books

Dees, Sarah. *The Big Book of Amazing LEGO Creations with Bricks You Already Have*. Salem, MA: Page Street Publishing Co., 2021.

Galvez-Aranda, Diego and Mauricio Galvez Legua. *Robotics Models Using LEGO WeDo 2.0*. New York, NY: Apress, 2021.

McCue, Camille. *Getting Started with Coding: Get Creative with Code!* Hoboken, NJ: Wiley, 2020.

Websites

Code.org
www.code.org
An educational nonprofit focused on providing computer science tools for grades K–12.

LEGO Education
www.education.lego.com
This is the LEGO Group's official website for all things learning with LEGO.

Scratch
www.scratch.mit.edu
This is the official website for the Scratch programming language.

Publisher's note to educators and parents: *Our editors have carefully reviewed these websites to ensure that they are suitable for students. Many websites change frequently, however, and we cannot guarantee that a site's future contents will continue to meet our high standards of quality and educational value. Be advised that students should be closely supervised whenever they access the internet.*

INDEX